Water Harvesters

Jenna Myers

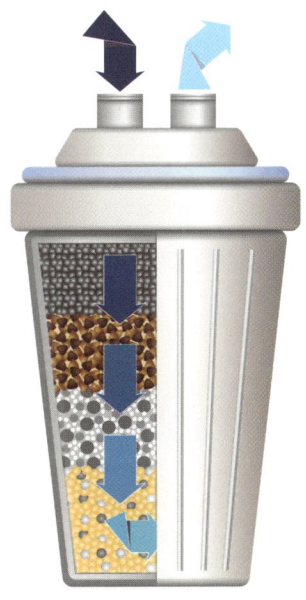

Series Editor **Casey Malarcher**

Level 3 - ❸

Water Harvesters

Jenna Myers

© 2018 Seed Learning, Inc.

Series Editor: Casey Malarcher
Acquisitions Editor: Anne Taylor
Copy Editor: Liana Robinson
Cover/Interior Design: Highline Studio

ISBN: 978-1-943980-45-1

10 9 8 7 6 5 4 3 2 1
22 21 20 19 18

Photo Credits

Contents

The Job of a Water Harvester

Water is an important natural resource. Most people take it for granted. We need water for life, yet we often waste it. We take long showers or leave the water running while we brush our teeth. Although water covers 70% of our planet, there will not always be so much. In fact, fresh water is incredibly rare.

Using water every day

By 2025, two-thirds of the world's population will more than likely suffer from water shortages. Due to these water stresses, we must find ways of collecting and using water before it gets wasted. This is where the job of a water harvester becomes essential for our future.

Storing water ▶

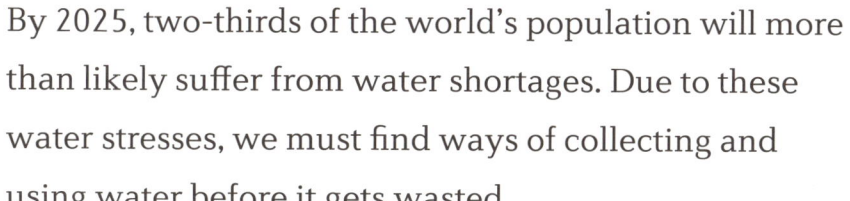

A short supply of water

In general, a water harvester is someone who collects and stores water. Water harvesters can capture rainwater. They can also get water from fog or air.

◀ A rain barrel

Built for collecting ▶
rainwater

A water harvester understands the environmental and economic benefits of harvesting water. The collected water can be used for gardens, grass, fields, and animals. It can even be used as drinking water once properly treated.

Water for drinking ▶

How to Be a Water Harvester

Rainwater harvesting is an ancient technique. It is known to have existed over 4,000 years ago. But there are other ways to harvest water, not just from rain. Classes for water harvesters teach these other ways. Those who take water harvesting courses are often already trained as plumbers or installers.

A plumber

A water-harvesting project

Some people in water harvesting courses are skill seekers. They are looking to further their career in the building trades. Others wish to learn more about water harvesting in order to use it in a large project. Such projects often focus on helping developing countries find an independent water supply.

Water harvesting systems can be simple or complex. Some systems can be put in with basic skills. Others use complicated computer systems that require advanced training.

A basic rainwater harvesting system is simple. It does not require a large amount of additional training to successfully build one. However, a more complex rainwater harvesting system might require the use of digital tools and technical skills.

Using a smartphone with ▶
a rainwater harvesting
system

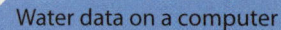

Water data on a computer

A water harvester will require a different set of skills and knowledge depending on what area he or she wishes to focus on.

It is important for those who want to make a career out of harvesting water to get proper training. Courses usually provide detailed instruction on water harvesting systems, planning, design, and installation.

Thinking about choices ▶

A drought

This training might include learning about ways to combine water-harvesting systems with other designs, such as wind or solar energy, farming, and animal habitats.

A house plan ▶

Solar panels and wind turbines

A water system for a house and a field

The training could also include classes on the application of sustainable living design, details about storage of collected water, and ways to match water-harvesting systems with appropriate landscaping.

Water Harvesters at Work

Rain is a great source of water, and water harvesters find collecting it convenient. Water harvesters capture water from surfaces on which rain falls. The rainwater that is collected can then be stored for later use. Most often rainwater is collected from the roofs of houses and buildings. It can also be collected from rock surfaces or hills.

Rain falling on a roof ▶

Collecting rainwater

Storage tanks

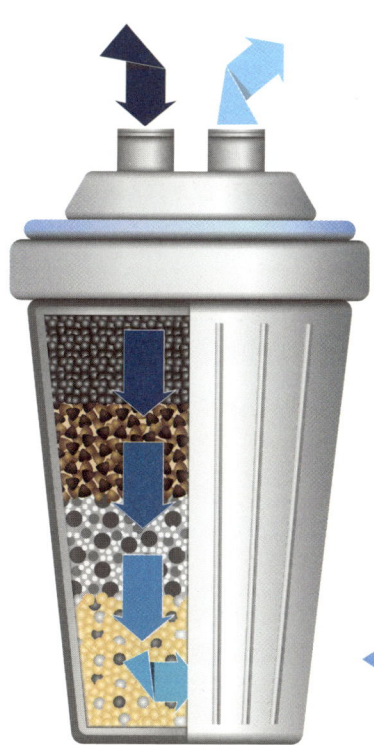

To capture the water, the water harvester must direct the flow of rainwater. The harvester must get the water to a storage tank. From there, the water must be filtered. Once filtered, it is then safe for drinking.

◀ A filter system

Fog in the desert

Due to the lack of rain in some areas of the world, it is not always possible to collect rainwater, store it, and use it. As a result, some water harvesters collect water from fog.

Fog and mist in the jungle

In some regions of the world, large nets are hung up in areas with thick fog and high winds. The wind pushes the fog through the net. As the fog condenses into water, it runs down the net. This water gets collected by water tanks or "fog catchers." This way of capturing water is not only cheap but also very effective.

Working together to save water

Water shortages in parts of the world where there is a lack of rain are still a big problem. However, when the weather is right and the maintenance costs are met, collecting water from fog has provided a solution for many at-risk communities around the world. It is a proven sustainable solution to the water issues in many places.

As the world's population continues to grow, water shortages will occur more often. While "fog harvesting" has been a successful way of capturing water in some countries, it does not work everywhere. Unfortunately, this method requires very special weather conditions.

A growing population ▶

A new technology developed by scientists can provide a way for water harvesters to get clean, fresh water almost anywhere on earth. This method draws water directly from the air, even in the driest parts of the world, such as deserts. This type of water harvesting uses a solar-powered device.

Developing new technology

500ml
15oz
450ml
400ml
350ml
10oz
300ml
250ml
200ml
5oz
150ml
100ml

Liquide
Liquide (liquid)
Liquide

Farine
Farina
Farine
Flour

300g
250g
200g
150g
100g

◄ Liquid water

In very dry conditions, this new method can provide 2.8 liters of water from the air over a 12-hour period. It does this by trapping water vapor. Solar power then drives the water toward a cooler condensing plate, which returns the vapor to liquid water so it can run into the device's collector.

A solar- ▶
powered
device

Looking to the Future

The growing need for water harvesting is creating new jobs. If you think about the number of roofs and roads without water harvesting systems, it is easy to understand the economic strength of water harvesting. Water harvesting projects are needed in both urban and rural areas. This means there will be more jobs than there will be water harvesters.

Most roofs lack water harvesting systems.

Rainwater harvesting is essential. This is because surface water (water on the surface of the earth such as in a river, lake, or ocean) and ground water (water that is below the ground) are not enough to meet the demands of the world's population.

Water on the surface of the earth

WATER IS PRECIOUS

SAVE IT

DO YOU HARVEST RAINWATER?

LOVE EARTH

More development of new technology is required to get the most benefits out of water harvesting. Therefore, there is a growing need for trained water harvesters who can build systems as well as do research.

Collecting rain as it falls

Perhaps in the future, most countries will use water harvesting as a cheap and safe source of clean water. Maybe one day the world will have its water shortage issues solved. Until then, those who choose to become water harvesters will find plenty of opportunities for work around the world.

Comprehension Questions

1. Due to the world's water stresses, we must . . .
   (a) build more solar-powered devices.
   (b) look for jobs in rural areas.
   (c) find ways of collecting and storing water.
   (d) provide detailed instruction on water harvesting systems.

2. Where do water harvesters get water from?
   (a) Air
   (b) Rain
   (c) Fog
   (d) All of the above

3. What is true about rainwater harvesting?
   (a) It is only possible using ground water.
   (b) It is a cheap and safe source of water.
   (c) It is a modern technique.
   (d) It can only be done in urban areas.

4. Which method of capturing water requires a net and windy weather?
   (a) Fog harvesting
   (b) Rainwater harvesting
   (c) Desert air harvesting
   (d) All of the above

5. The solar-powered water harvester can provide 2.8 liters of water from the air over a 12-hour period by . . .
   (a) heating very cold water.
   (b) collecting water from fog.
   (c) trapping water vapor.
   (d) catching rainwater.

Glossary

- **application** (n.) the practical use of an idea or method

- **capture** (v.) to get something and keep it for a particular reason

- **condense** (v.) to change from a gas or vapor into a liquid

- **digital** (adj.) using or characterized by computer technology

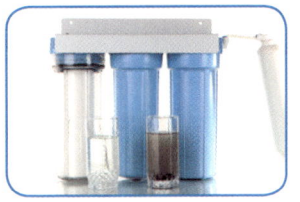

- **filter** (v.) to pass liquid through a device that is used to remove something unwanted from it

- **fog** (n.) a thick cloud of very small drops of water in the air close to the land or sea

- **habitat** (n.) the place or type of place where a plant or animal naturally or normally lives or grows

- **installer** (n.) a person who places or fixes equipment or machinery in position ready for use

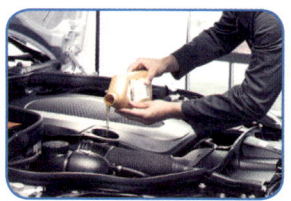

- **maintenance** (n.) the act of keeping property or equipment in good condition by making repairs and correcting problems

- **net** (n.) many strong threads or ropes that are put together in a way to be used to catch or hold things

- **plumber** (n.) a person whose job is to install or repair sinks, toilets, water pipes, etc.

- **rainwater** (n.) water that falls as rain

- **rural** (adj.) relating to the countryside and not to towns

- **shortage** (n.) a situation when there is not enough of the people or things that are needed

- **sustainable** (adj.) causing little or no damage to the environment and able to continue for a long time

- **take for granted** (v. phr.) to expect something and not understand that you are lucky to have it

- **urban** (adj.) relating to a city or town and not to the countryside

- **vapor** (n.) a mass of very small drops of liquid in the air

Notes

Here are some tools and ideas that water harvesters are using or developing to provide the world with more sources of clean water. Readers may enjoy researching these topics to learn more about this field.

GIS (Geographic Information Systems) tools can help people identify the rainwater harvesting potential in a specific area. It is important to understand exactly where rainwater is best harvested to make full use of its potential.

Artificial lakes are man-made lakes that catch rainwater or river water. They can be small for private use or very large for public use. The stored water can be used for irrigation, drinking water (after purification) or energy production. These lakes can also provide drinking water for wild animals.

Storm water harvesting involves collecting, treating, storing and using storm water runoff from urban areas.

Be a water harvester yourself. All you need is a way to catch rainwater, a container to store it, and some ideas of how to use it where it is most needed.

List of Books

LEVEL 1

1. Robotics Engineers
2. Cyber Security Experts
3. Medical Scientists
4. Social Media Managers
5. Asset Managers

LEVEL 2

1. Drone Pilots
2. App Developers
3. Wearable Technology Creators
4. Computer Intelligence Engineers
5. Digital Modelers

LEVEL 3

1. IoT Marketing Specialists
2. Space Pilots
3. Water Harvesters
4. Genetic Counselors
5. Data Miners

LEVEL 4

1. Database Administrators
2. Nanotechnology Research Scientists
3. Quantum Computer Scientists
4. Agricultural Engineers
5. Intellectual Property Lawyers

"The future of the economy is in STEM. That's where the jobs of tomorrow will be."

James Brown (Executive Director of the STEM Education Coalition in Washington, D.C.)

Data from the US Bureau of Labor Statistics (BLS) support that assertion. Employment in occupations related to STEM—science, technology, engineering, and mathematics—is projected to grow to more than 9 million by 2022 [in the US alone] ... Overall, STEM occupations are projected to grow faster than the average for all occupations.

from *STEM 101: Intro to Tomorrow's Jobs* US Bureau of Labor Statistics